CIA MANUAL FOR PSYCHOLOGICAL OPERATIONS IN GUERRILLA WARFARE

CIA MANUAL FOR PSYCHO-LOGICAL OPERATIONS IN GUERRILLA WARFARE

DUANE RAMSDELL "DEWEY" CLARRIDGE

A tactical manual for the revolutionary. First published by the Central Intelligence Agency and distributed to the Contras in Central America.

Republished by
www.BIGFONTBOOKS.com

ISBN: 978-1-963956-20-7

Contents

Foreword from the Editor   7
PREFACE   11
INTRODUCTION   13
COMBATANT-PROPAGANDIST GUERRILLA   17
ARMED PROPAGANDA   23
ARMED PROPAGANDA TEAMS (APTs)   29
DEVELOPMENT AND CONTROL OF
FRONT ORGANIZATIONS   39
CONTROL OF MASS CONCENTRATIONS
AND MEETINGS   45
MASSIVE IN-DEPTH SUPPORT
THROUGH PSYCHOLOGICAL OPERATIONS   51
APPENDIX   53

Foreword from the Editor

This booklet was designed for the contras fighting the Socialist movement of the Nicaraguan Sandinista movement. The CIA had attempted to borrow from the long tradition of underground warfare and psychological operations that were successful throughout history against, usually a much more powerful adversary.

Guerrilla warfare is a form of irregular warfare and refers to conflicts in which a small group of combatants including, but not limited to, armed civilians (or "irregulars") use military tactics, such as ambushes,sabotage, raids, the element of surprise, and extraordinary mobility to harass a larger and less-mobile traditional army, or strike a vulnerable target, and withdraw almost immediately.

The term means "little war" in Spanish, and the word, guerrilla, has been used to describe the concept since the 18th century, and perhaps earlier.

The tactics of guerrilla warfare were used successfully in the 20th century by—among others— Mao Zedong and the People's Liberation Army in the Second Sino-Japanese War and Chinese Civil War, Fidel Castro,Che Guevara and the 26th of July Movement in the Cuban Revolution, Ho Chi Minh, Vo Nguyen Giap, Viet Cong and select members of the Green Berets in the Vietnam War, the Liberation Tigers of Tamil Eelam in the Sri Lankan Civil War, the Afghan Mujahideen in the Soviet war in Afghanistan,George Grivas and Nikos Sampson's Greek guerrilla group EOKA in Cyprus, Aris Velouchiotis and Stefanos Sarafis and the EAM against theAxis occupation of Greece during World War II, Paul Emil von Lettow-Vorbeck and the GermanSchutztruppe in World War I, Josip Broz Tito and the Yugoslav Partisans in World War II, and the antifrancoist guerrilla in Spain during the Franco dictatorship, the Kosovo Liberation Army in the Kosovo War, and the Irish Republican Army during the Irish War of Independence. Most factions of the Taliban, Iraqi Insurgency, Colombia's FARC, and the Communist Party of India (Maoist) are said to be engaged in some form of guerrilla warfare—as was, until recently, the Communist Party of Nepal (Maoist). In India, Marathas under leadership of Shivaji used it to overthrow of the Mughals. It was also effectively used by Tatya Tope and Rani Laxmibai in the Indian Rebellion of 1857, as well as byPazhassi Raja of Kerala to fight the British.

In War and Peace (written in 1865-1869, in part about Napoleon's invasion of Russia), Leo Tolstoy says that guerrilla warfare is named after the Guerrillas in Spain. He appears to be referring to a specific group that used guerrilla warfare in a war fought in Spain before the 1860s.

That war began in 1808 with the occupation of Spain by Napoleon's French army. Bands of guerrillas (so named; one of the most important led by Juan Martin Diez, Agustina de Aragón orJuana La Galana) and the normal Spanish army both fought Napoleon. Our modern word "guerrilla" traces its origin to these bands in this war. These guerrillas were very effective in fighting Napoleon. Their principal function was to disrupt the supply and communication lines of the French army by intercepting messages and by seizing convoys of supplies, arms, and money. They did so much damage to Napoleon's army that Joseph Leopold Hugo, a French general, was ordered to "pursue exclusively" Diez and his guerrillas. According to Merriam Webster's Collegiate Dictionary, the word "guerrilla" was first used as a noun in 1809 and as an adjective in 1811.

An early example of this came when General John Burgoyne who, during the Saratoga campaign of the American War of Independence, noted that, in proceeding through dense woodland:

"The enemy is infinitely inferior to the King's Troop in open space, and hardy combat, is well fitted by disposition and practice, for the stratagems of enterprises of Little War...upon the same principle must be a constant rule, in or near woods to place advanced sentries, where they may have a tree or some other defence to prevent their being taken off by a single marksman."

Sun Tzu, in his The Art of War (6th century BCE), was the earliest to propose the use of guerrilla warfare. This directly inspired the development of modern guerrilla warfare. Communist leaders like Mao Zedong and North Vietnamese Ho Chi Minh both implemented guerrilla warfare in the style of Sun Tzu, which served as a model for similar strategies elsewhere, such as the Cuban "foco" theory and the anti-Soviet Mujahadeen in Afghanistan. While the tactics of modern guerrilla warfare originate in the 20th century, irregular warfare, using elements later characteristic of modern guerrilla warfare, has existed throughout the battles of many ancient civilizations.

The influence of the ancient military philosopher Sun Tzu on Mao's military thought will be apparent to those who have read The Book of War. Sun Tzu wrote that speed, surprise, and deception were the primary essentials of the attack, and his succinct advice: Sheng Tung, Chi Hsi (Distraction in the East, Strike in the West) is no less valid today than it was when he wrote it 2,400 years ago...Guerrilla tactical doctrine may be summarized in four Chinese characters pronounced "Sheng Tung, Chi Hsi," which mean "Uproar [in the] East; Strike [in the] West". Here we find expressed the all- important principles of distraction on the one hand and concentration on the other; to fix the enemy's attention and to strike where and when he least anticipates the blow.—Samuel B. Griffith, On Guerrilla War

The earliest description of guerrilla warfare is an alleged battle between Emperor Huang and the Miao in China. Guerrilla warfare was not unique to China, nomadic and migratory tribes such as the Scythians, Goths, Huns, and Magyars used elements of guerrilla warfare to fight the Persian Empire,Roman Empire, and Alexander the Great.

The Mongols faced irregulars composed of armed peasants in Hungary after the Battle of Mohi. The various castles provided power bases for the Hungarian resistance fighters; while the Mongols devastated the countryside, the Mongols were unable to take the castles and walled cities. In 1242, the Hungarians ambushed and destroyed two toumens of rearguard troops in the Carpathian mountains, where light horse is at a disadvantage because of rough terrain.

PREFACE

Guerrilla warfare is essentially a political war. Therefore, its area of operations exceeds the territorial limits of conventional warfare, to penetrate the political entity itself: the "political animal" that Aristotle defined.

In effect, the human being should be considered the priority objective in a political war. And conceived as the military target of guerrilla war, the human being has his most critical point in his mind. Once his mind has been reached, the "political animal" has been defeated, without necessarily receiving bullets.

Guerrilla warfare is born and grows in the political environment; in the constant combat to dominate that area of political mentality that is inherent to all human beings and which collectively constitutes the "environment" in which guerrilla warfare moves, and which is where precisely its victory or failure is defined.

This conception of guerrilla warfare as political war turns Psychological Operations into the decisive factor of the results. The target, then, is the minds of the population, all the population: our troops, the enemy troops and the civilian population.

This book is a manual for the training of guerrillas in psychological operations, and its application to the concrete case of the Christian and democratic crusade being waged in Nicaragua by the Freedom Commandos.

Note: In this text the editor has replaced the phrase Sandinista with "Government" as this manual should be a guide to fight and conquer any oppressive government.

INTRODUCTION

Generalities

The purpose of this book is to introduce the guerrilla student to the psychological operations techniques that will be of immediate and practical value to him in guerrilla warfare. This section is introductory and general; subsequent sections will cover each point set forth here in more detail.

The nature of the environment of guerrilla warfare does not permit sophisticated psychological operations, and it becomes necessary for the chiefs of groups, chiefs of detachments and squadron leaders to have the ability to carry out, with minimal instructions from the higher levels, psychological action operations with the contacts that are thoroughly aware of the situation, i.e. the foundations.

Combatant-Propagandist Guerrillas

In order to obtain the maximum results from the psychological operations in guerrilla warfare, every combatant should be as highly motivated to carry out propaganda face to face as he is a combatant. This means that the individual political awareness of the guerrilla of the reason for his struggle will be as acute as his ability to fight.

Such a political awareness and motivation is obtained through the dynamic of groups and self-criticism, as a standard method of instruction for the guerrilla training and operations. Group discussions raise the spirit and improve the unity of thought of the guerrilla training and operations. Group discussions raise the spirit and improve the unity of thought of the guerrilla squads and exercise social pressure on the weak members to carry out a better role in future training or in combative action. Self-criticism is in terms of one's contribution or defects in his contribution to the cause, to the movement, the struggle, etc.; and gives a positive individual commitment to the mission of the group.

The desired result is a guerrilla who can persuasively justify his actions when he comes into contact with any member of the People of Nicaragua, and especially with himself and his fellow guerrillas in dealing with the vicissitudes of guerrilla warfare. This means that every guerrilla will be persuasive in his face-to-face communication - propagandist-combatant - is his contact with the people; he should be able to give 5 or 10 logical reasons why, for example, a peasant should give him cloth, needle and thread to mend his clothes. When the guerrilla behaves in this manner, enemy propaganda will never succeed in making him an enemy in the eyes of the people. It also means that hunger, cold, fatigue and insecurity will have a meaning, psychologically, in the cause of the struggle due to his constant orientation.

Armed Propaganda

Armed propaganda includes every act carried out, and the good impression that this armed force causes will result in positive attitudes in the population toward that force; ad it does not include forced indoctrination. Armed propaganda improves the behavior of the population toward them, and it is not achieved by force.

This means that a guerilla-armed unit in a rural town will not give the impression that arms are their strength over the peasants, but rather that they are the strength of the peasants against the government of repression. This is achieved through a close identification with the people, as follows: hanging up weapons and working together with them on their crops, in construction, in the harvesting of grains, in fishing, etc.; explanations to young men about basic weapons, e.g. giving them an unloaded weapon and letting them touch it, see it, etc.; describing in a rudimentary manner its operation; describing with simple slogans how weapons will serve the people to win their freedom; demanding the requests by the people for hospitals and education, reducing taxes, etc.

All these acts have as their goal the creation of an identification of the people with the weapons and the guerrillas who carry them, so that the population feels that the weapons are, indirectly, their weapon to protect them and help them in the struggle against a regime of oppression. Implicit terror always accompanies weapons, since the people are internally "aware" that they can be used against them, but as long as explicit coercion is avoided, positive attitudes can be achieved with respect to the presence of armed guerrillas within the population.

Armed Propaganda Teams

Armed Propaganda Teams (EPA) are formed through a careful selection of persuasive and highly motivated guerrillas who move about within the population, encouraging the people to support the guerrillas and put up resistance against the enemy. It combines a high degree of political awareness and the "armed" propaganda ability of the guerrillas toward a planned, programmed, and controlled effort.

The careful selection of the staff, based on their persuasiveness in informal discussions and their ability in combat, is more important than their degree of education or the training program. The tactics of the Armed Propaganda Teams are carried out covertly, and should be parallel to the tactical effort in guerrilla warfare. The knowledge of the psychology of the population is primary for the Armed Propaganda Teams, but much more intelligence data will be obtained from an EPA program in the area of operations.

Development and Control of the "Front" Organizations

The development and control of "front" (or facade) organizations is carried out through subjective internal control at group meetings of "inside cadres," and the calculations of the time for the fusion of these combined efforts to be applied to the masses.

Established citizens-doctors, lawyers, businessmen, teachers, etc. - will be recruited initially as "Social Crusaders" in typically "innocuous" movements in the area of operations. When their "involvement" with the clandestine organization is revealed to them, this supplies the psychological pressure to use them as "inside cadres" in groups to which they already belong or of which they can be members.

Then they will receive instruction in techniques of persuasion over control of target groups to support our democratic revolution, through a gradual and skillful process. A cell control system isolates individuals from one another, and at the appropriate moment, their influence is used for the fusion of groups in a united national front.

Control of Meetings and Mass Assemblies

The control of mass meetings in support of guerrilla warfare is carried out internally through a covert commando element, bodyguards, messengers, shock forces (initiators of incidents), placard carriers (also used for making signals), shouters of slogans, everything under the control of the outside commando element.

When the cadres are placed or recruited in organizations such as labor unions, youth groups agrarian organizations or professional associations, they will begin to manipulate the objectives of the groups. The psychological apparatus of our movement through inside cadres prepares a mental attitude which at the crucial moment can be turned into a fury of justified violence.

Through a small group of guerrillas infiltrated within the masses this can be carried out; they will have the mission of agitating by giving the impression that there are many of them and that they have a large popular backing. Using the tactics of a force of 200-300 agitators, a demonstration can be created in which 10,000-20,000 persons take part.

The support of local contacts who are familiar with the deep reality is achieved through the exploitation of the social and political weaknesses of the target society, with propagandist-combatant guerrillas, armed propaganda, armed propaganda teams, cover organizations and mass meetings.

The combatant-propagandist guerrilla is the result of a continuous program of indoctrination and motivation. They will have the mission of showing the people how great and fair our movement is in the eyes of all Nicaraguans and the world. Identifying themselves with our people, they will increase the sympathy towards our movement, which will result in greater support of the population for the freedom commandos, taking away support for the regime in power.

Armed propaganda will extend this identification process of the people with the Christian guerrillas, providing converging points against the regime.

The Armed Propaganda Teams provide a several-stage program of persuasive planning in guerrilla warfare in all areas of the country. Also, these teams are the "eyes and ears" of our movement.

The development and control of the cover organizations in guerrilla warfare will give our movement the ability to create a "whiplash" effect within the population when the order for fusion is given. When the infiltration and internal subjective control have been developed in a manner parallel to other guerrilla activities, a commandant of ours will literally be able to shake up the Government structure, and replace it.

The mass assemblies and meetings are the culmination of a wide base support among the population, and it comes about in the later phases of the operation. This is the moment in which the overthrow can be achieved and our revolution can become an open one, requiring the close collaboration of the entire population of the country, and of contacts with their roots in reality.

The tactical effort in guerrilla warfare is directed at the weaknesses of the enemy and at destroying their military resistance capacity, and should be parallel to a psychological effort to weaken and destroy their sociopolitical capacity at the same time. In guerrilla warfare, more than in any other type of military effort, the psychological activities should be simultaneous with the military ones, in order to achieve the objectives desired.

COMBATANT-PROPAGANDIST GUERRILLA

Generalities

The objective of this section is to familiarize the guerrilla with the techniques of psychological operations, which maximizes the social-psychological effect of a guerrilla movement, converting the guerrilla into a propagandist, in addition to being a combatant. The nature of the environment in guerrilla warfare does not permit sophisticated facilities for psychological operations, so that use should be made of the effective face-to-face persuasion of each guerrilla.

Political Awareness

The individual political awareness of the guerrilla, the reason for his struggle, will be as important as his ability in combat. This political awareness and motivation will be achieved:

By improving the combat potential of the guerrilla by improving his motivation for fighting.

By the guerrilla recognizing himself as a vital tie between the democratic guerrillas and the people, whose support is essential for the subsistence of both.

By fostering the support of the population for the national insurgence through the support for the guerrillas of the locale, which provides a psychological basis in the population for politics after the victory has been achieved.

By developing trust in the guerrillas and in the population, for the reconstruction of a local and national government.

By promoting the value of participation by the guerrillas and the people in the civic affairs of the insurrection and in the national programs.

By developing in each guerrilla the ability of persuasion face-to-face, at the local level, to win the support of the population, which is essential for success in guerrilla warfare.

Group Dynamics

This political awareness building and motivation are attained by the use of group dynamics at the level of small units. The group discussion method and self-criticism are a general guerrilla training and operations technique.

Group discussions raise the spirit and increase the unity of thought in small guerrilla groups and exercise social pressure on the weakest members to better carry out their mission in training and future combat actions. These group discussions will give special emphasis to:

Creating a favorable opinion of our movement. Through local and national history, make it clear that the regime is "foreignizing," "repressive" and "imperialistic," and that even though there are some Nicaraguans within the government, point out that they are "puppets" of the power of the Soviets and Cubans, i.e. of foreign power.

Always a local focus. Matters of an international nature will be explained only in support of local events in the guerrilla warfare.

The unification of the nation is our goal. This means that the defeat of the Government armed forces is our priority. Our insurrectional movement is a pluralistic political platform from which we are determined to win freedom, equality, a better economy with work facilities, a higher standard of living, a true democracy for all Nicaraguans without exception.

Providing to each guerrilla clear understanding about the struggle for national sovereignty against Soviet-Cuban imperialism. Discussion guides will lead the guerrillas so that they will see the injustices of the Government system.

Showing each guerrilla the need for good behavior to win the support of the population. Discussion guides should convince the guerrillas that the attitude and opinion of the population play a decisive role, because victory is impossible without popular support.

Self-criticism will be in constructive terms that will contribute to the mission of the movement, and which will provide the guerrillas with the conviction that they have a constant and positive individual responsibility in the mission of the group. The method of instruction will be:

Divisions of the guerrilla force into squads for group discussions, including command and support elements, whenever the tactical situation permits it. The makeup of the small units should be maintained when these groups are designated.

Assignment of a political cadre in the guerrilla force to each group to guide the discussion. The squad leader should help the cadre to foster study and the expression of thoughts. If there are not enough political cadres for each squad or post, leaders should guide the discussions, and the available cadres visit alternate groups.

It is appropriate for the cadre (or the leader) to guide the discussion of a group to cover a number of points and to reach a correct conclusion. The guerrillas should feel that it was their free and own decision. The cadre should serve as a private teacher. The cadre or leader will not act as a lecturer, but will help the members of the group to study and express their own opinions.

The political cadre will at the end of every discussion make a summary of the principal points, leading them to the correct conclusions. Any serious difference with the objectives of the movement should be noted by the cadre and reported to the commandant of the force. If necessary, a combined group meeting will be held and the team of political cadres will explain and rectify the misunderstanding.

Democratic conduct by the political cadres: living, eating and working with the guerrillas, and if possible, fighting at their side, sharing their living conditions. All of this will foster understanding and the spirit of cooperation that will help in the discussion and exchange of ideas.

Carry out-group discussions in towns, and areas of operations whenever possible with the civilian population, and not limit them to camps or bases. This is done to emphasize the revolutionary nature of the struggle and to demonstrate that the guerrillas identified with the objectives of the people move about within the population. The guerrilla projects himself toward the people, as the political cadre does toward the guerrilla, and they should live, eat and work together to realize a unity of revolutionary thought.

The principles for guerrilla and political-cadre group discussions are:

Organize discussion groups at the post or squad level. A cadre cannot be sure of the comprehension and acceptance of the concepts and conclusions by guerrillas in large groups. In a group of the size of a squad of 10 men, the judgment and control of the situation is greater. In this way, all students will participate in an exchange among them; the political leader, the group leader, and also the political cadre. Special attention will be given to the individual ability to discuss the objectives of the insurrectional struggle. Whenever a guerrilla expresses his opinion, he will be interested in listening to the opinions of others, leading as a result to the unity of thought.

Combine the different points of view and reach an opinion or common conclusion. This is the most difficult task of a political guerrilla cadre. After the group discussions of the democratic objectives of the movement, the chief of the team of political cadres of the guerrilla force should combine the conclusions of individual groups in a general summary. At a meeting with all the discussion groups, the cadre shall provide the principal points, and the guerrillas will have the opportunity to clarify or modify their points of view. To carry this out, the conclusions will be summarized in the form of slogans, wherever possible.

Face with honesty the national and local problems of our struggle. The political cadres should always be prepared to discuss solutions to the problems observed by the guerrillas. During the discussions, the guerrillas should be guided by the following three principles:

Freedom of thought.
Freedom of expression.

Concentration of thoughts on the objectives of the democratic struggle.

The result desired is a guerrilla who in a persuasive manner ca justify all of his acts whenever he is in contact with any member of the town/people and especially with himself and with his guerrilla companion by facing the vicissitudes of guerrilla warfare.

This means that every guerrilla will come to have effective face-to face persuasion as a combatant-propagandist in his contact with the people to the point of giving 5-10 logical reasons why, e.g. a peasant should give him a piece of cloth, or a needle and thread to mend his clothes. When behaves in this manner, no type of propaganda of the enemy will be able to make a "terrorist" of him in the eyes of the people.

In addition, hunger, cold, fatigue and insecurity in the existence of the guerrilla acquire meaning in the cause of the struggle due to the constant psychological orientation.

Camp Procedures

Encamping the guerrilla units gives greater motivation, in addition to reducing distractions, and increases the spirit of cooperation of small units, relating the physical environment to the psychological one. The squad chief shall establish the regular camping procedure. Once thy have divested themselves of their packs, the chief will choose the appropriate ground for camping. He should select land that predominates over the zone with two or three escape routes. He will choose among his men and give them responsibilities such as:

Clean the camp area.

Provide adequate drainage in case of rain. Also build some trenches or holes for marksmen in case of emergency. In addition, he will build a stove, which will be done by making some small trenches and placing three rocks in place; in case the stove is built on a pedestal, it will be filled with clay and rocks.

Build a wind-breaking wall, which will be covered on the sides and on the top with branches and leaves of the same vegetation of the zones. This will serve for camouflaging and protecting it from aerial visibility or from enemy patrols around.

Construct a latrine and a hole where waste and garbage will be buried, which should be covered over at the time of abandoning the camp.

Once the camp has been set up, it is recommended that a watchman be positioned in the places of access at a prudent distance, where the shout of alarm can be heard. In the same moment the password will be established, which should be changed every 24 hours. The commander should establish ahead of time an alternate meeting point, in case of having to abandon the camp in a hurried manner, and they will be able to meet in the other already established point, and they should warn the patrol that if at a particular time they cannot meet at the established point, the should have a third meeting point.

These procedures contribute to the motivation of the guerrilla and improve the spirit of cooperation in the unit. The danger, sense of insecurity, anxiety and daily concern in the life of a guerrilla require tangible evidence of belonging in an order for him to keep up his spirit and morale.

In addition to the good physical conditions in which the guerrilla should find himself, good psychological conditions are necessary, for which group discussions and becoming a self-critic are recommended, which will greatly benefit the spirit and morale of the same.

Having broken camp with the effort and cooperation of everyone strengthens the spirit of the group. The guerrilla will be inclined then towards the unity of thought in democratic objectives.

Interaction with the People

In order to ensure popular support, essential for the good development of guerrilla warfare, the leaders should induce a positive interaction between the civilians and the guerrillas, through the principle of "live, eat , and work with the people," and maintain control of their activities. In group discussions, the leaders and political cadres should give emphasis to positively identifying themselves with the people.

It is not recommendable to speak of military tactical plans in discussions with civilians. The Communist foe should be pointed out as the number one enemy of the people, and as a secondary threat against our guerrilla forces.

Whenever there is a chance, groups of members should be chosen who have a high political awareness and high disciplinary conduct in the work to be carried out, in order to be sent to the populous areas in order to direct the armed propaganda, where they should persuade the people through dialogue in face-to-face confrontations, where these principles should be followed:

Respect for human rights and others' property.

Helping the people in community work.

Protecting the people from Communist aggressions.

Teaching the people environmental hygiene, to read, etc., in order to win their trust, which will lead to a better democratic ideological preparation.

This attitude will foster the sympathy of the peasants for our movement, and they will immediately become one of us, through logistical support, coverage and intelligence information on the enemy or participation in combat. The guerrillas should be persuasive through the word and not dictatorial with weapons. If they behave in this way, the people will feel respected, will be more inclined to accept our message and will consolidate into popular support.

In any place in which tactical guerrilla operations are carried out in populous areas, the squad should undertake psychological actions parallel to these, and should proceed, accompany and consolidate the common objective and explain to all the people about our struggle, explaining that our presence is to give peace, liberty and democracy to all Nicaraguans without exception, and explaining that out struggle is not against the nationals but rather against Russian imperialism. This will serve to ensure greater Psychological achievements which will increase the operations of the future.

Conclusions

The nature of the environment in guerrilla warfare does not permit sophisticated facilities for psychological operations, and the face-to-face persuasion of the guerrilla combatant-propagandists with the people is an effective and available tool which we should use as much as possible during the process of the struggle.

ARMED PROPAGANDA

Generalities

Frequently a misunderstanding exists on "armed propaganda," that this tactic is a compulsion of the people with arms. In reality, it does not include compulsion, but the guerrilla should know well the principles and methods of this tactic. The objective of this section is to give the guerrilla student an understanding of the armed propaganda that should be used, and that will be able to be applied in guerrilla warfare.

Close Identification with the People

Armed propaganda includes all acts carried out by an armed force, whose results improve the attitude of the people toward this force, and it does not include forced indoctrination. This is carried out by a close identification with the people on any occasion. For example:

Putting aside weapons and working side by side with the peasants in the countryside: building, fishing, repairing roofs, transporting water, etc.

When working with the people, the guerrillas can use slogans such as "many hands doing small things, but doing them together."

Participating in the tasks of the people, they can establish a strong tie between them and the guerrillas and at the same time a popular support for our movement is generated.

During the patrols and other operations around or in the midst of villages, each guerrilla should be respectful and courteous with the people. In addition he should move with care and always be well prepared to fight, if necessary. But he should not always see all the people as enemies, with suspicions or hostility. Even in war, it is possible to smile, laugh or greet people. Truly, the cause of our revolutionary base, the reason why we are struggling, is our people. We must be respectful to them on all occasions that present themselves.

In places and situations wherever possible, e.g. when they are resting during the march, the guerrillas can explain the operation of weapons to the youths and young men. They can show them an unloaded rifle so that they will learn to load it and unload it; their use, and aiming at imaginary targets they are potential recruits for our forces.

The guerrillas should always be prepared with simple slogans in order to explain to the people, whether in an intentional form or by chance, the reason for the weapons.

"The weapons will be for winning freedom; they are for you."

"With weapons we can impose demands such as hospitals, schools, better roads, and social services for the people, for you."

"Our weapons are, in truth, the weapons of the people, yours."

"With weapons we can change the Sandino-Communist regime and return to the people a true democracy so that we will all have economic opportunities."

All of this should be designed to create an identification of the people with the weapons and the guerrillas who carry them. Finally, we should make the people feel that we are thinking of them and that the weapons are the people's, in order to help them and protect them from a Communist, totalitarian, imperialist regime, indifferent to the needs of the population.

Implicit and Explicit Terror

A guerrilla-armed force always involves implicit terror because the population, without saying it aloud, feels terror that the weapons may be used against them. However, if the terror does not become explicit, positive results can be expected.

In a revolution, the individual lives under a constant threat of physical damage. If the government police cannot put an end to the guerrilla activities, the population will lose confidence in the government, which has the inherent mission of guaranteeing the safety of citizens. However, the guerrillas should be careful not to become an explicit terror, because this would result in a loss of popular support.

In the words of a leader of the Huk guerrilla movement of the Philippine Islands: "The population is always impressed by weapons, not by the terror that they cause, but rather by a sensation of strength/force. We must appear before the people, giving them the message of the struggle." This is, then, in a few words, the essence of armed propaganda.

An armed guerrilla force can occupy an entire town or small city that is neutral or relatively passive in the conflict. In order to conduct the armed propaganda in an effective manner, the following should be carried out simultaneously:

Destroy the military or police installations and remove the

survivors to a "public place."

Cut all the outside lines of communications: cables, radio, messengers. Set up ambushes in order to delay the reinforcements in all the possible entry routes.

Kidnap all officials or agents of the government and replace them in "public Places" with military or civilian persons of trust to our movement; in addition, carry out the following:

Establish a public tribunal that depends on the guerrillas, and cover the town or city in order to gather the population for this event.

Shame, ridicule and humiliate the "personal symbols" of the government of repression in the presence of the people and foster popular participation through guerrillas within the multitude, shouting slogans and jeers.

Reduce the influence of individuals in tune with the regime, pointing out their weaknesses and taking them out of the town, without damaging them publicly.

Mix the guerrillas within the population and show very good conduct by all members of the column, practicing the following:

Any article taken will be paid for with cash.

The hospitality offered by the people will be accepted and this opportunity will be exploited in order to carry out face-to-face persuasion about the struggle.

Courtesy visits should be made to the prominent persons and those with prestige in the place, such as doctors, priests, teachers, etc.

The guerrillas should instruct the population that with the end of the operative, and when the Government repressive forces interrogate them, they may reveal EVERYTHING about the military operation carried out. For example, the type of weapons they use, how many men arrived, from what direction they came and in what direction they left, in short, EVERYTHING.

In addition, indicate to the population that at meetings or in private discussion they can give the names of the Government informants, who will be removed together with the other officials of the government of repression.

When a meeting is held, conclude it with a speech by one of the leaders of guerrilla political cadres (the most dynamic), which includes explicit references to:

The fact that the "enemies of the people" -- the officials or Government agents -- must not be mistreated in spite of their criminal acts, although the guerrilla force may have suffered casualties, and that this is done due to the generosity of the Christian guerrillas.

Give a declaration of gratitude for the "hospitality" of the population, as well as let them know that the risks that they will run when the Governments return are greatly appreciated.

The fact that the regime, although it exploits the people with taxes, control of money, grains and all aspects of public life through associations, which they are forced to become part of, will not be able to resist the attacks of our guerrilla forces.

Make the promise to the people that you will return to ensure that the "leeches" of the regime of repression will not be able to hinder our guerrillas from integrating with the population.

A statement repeated to the population to the effect that they can reveal everything about this visit of our commandos, because we are not afraid of anything or anyone, neither the Soviets nor the Cubans. Emphasize that we are Nicaraguans, that we are fighting for the freedom of Nicaragua and to establish a very Nicaraguan government.

Guerrilla Weapons Are The Strength of the People over an Illegal Government

The armed propaganda in populated areas does not give the impression that weapons are the power of the guerrillas over the people, but

rather that the weapons are the strength of the people against a regime of repression. Whenever it is necessary to use armed force in an occupation or visit to a town or village, guerrillas should emphasize making sure that they:

Explain to the population that in the first place this is being done to protect them, the people, and not themselves.

Admit frankly and publicly that this is an "act of the democratic guerrilla movement," with appropriate explanations.

That this action, although it is not desirable, is necessary because the final objective of the insurrection is a free and democratic society, where acts of force are not necessary.

The force of weapons is a necessity caused by the oppressive system, and will cease to exist when the "forces of justice" of our movement assume control.

If, for example, it should be necessary for one of the advanced posts to have to fire on a citizen who was trying to leave the town or city in which the guerrillas are carrying out armed propaganda or political proselytism, the following is recommended:

Explain that if that citizen had managed to escape, he would have alerted the enemy that is near the town or city, and they could carry out acts of reprisal such as rapes, pillage, destruction, captures, etc., it this way terrorizing the inhabitants of the place for having given attention and hospitalities to the guerrillas of the town.

If a guerrilla fires at an individual, make the town see that he was an enemy of the people, and that they shot him because the guerrilla recognized as their first duty the protection of citizens.

The command tried to detain the informant without firing because he, like all Christian guerrillas, espouses nonviolence.

Firing at the Government informant, although it is against his own will, was necessary to prevent the repression of the government against innocent people.

Make the population see that it was the repressive system of the regime that was the cause of this situation, what really killed the informer, and that the weapon fired was one recovered in combat against the regime.

Make the population see that if the regime had ended the repression, the corruption backed by foreign powers, etc., the freedom commandos would not have had to brandish arms against brother Nicaraguans, which goes against our Christian sentiments. If the informant hadn't tried to escape he would be enjoying life together with the rest of the population, because not have tried to inform the enemy. This death would have been avoided if justice and freedom existed in Nicaragua, which is exactly the objective of the democratic guerrilla.

Selective Use of Violence for Propagandistic Effects

It is possible to neutralize carefully selected and planned targets, such as court judges, mesta judges, police and State Security officials, CDS chiefs, etc. For psychological purposes it is necessary to gather together the population affected, so that they will be present, take part in the act, and formulate accusations against the oppressor.

The target or person should be chosen on the basis of:

The spontaneous hostility that the majority of the population feels toward the target.

Use rejection or potential hatred by the majority of the population affected toward the target, stirring up the population and making them see all the negative and hostile actions of the individual against the people.

If the majority of the people give their support or backing to the target or subject, do not try to change these sentiments through provocation.

Relative difficulty of controlling the person who will replace the target.

The person who will replace the target should be chosen carefully, based on:

Degree of violence necessary to carry out the change.

Degree of violence acceptable to the population affected.

Degree of predictable reprisal by the enemy on the population affected or other individuals in the area of the target.

The mission to replace the individual should be followed by:

Extensive explanation within the population affected of the reason why it was necessary for the good of the people.

Explain that Government retaliation is unjust, indiscriminate, and above all, a justification for the execution of this mission.

Carefully test the reaction of the people toward the mission, as well as control this reaction, making sure that the populations reaction is beneficial towards the Freedom Commandos.

Conclusions

Armed propaganda includes all acts executed and the impact achieved by an armed force, which as a result produces positive attitudes in the population toward this force, and it does not include forced indoctrination. However, armed propaganda is the most effective available instrument of a guerrilla force.

ARMED PROPAGANDA TEAMS (APTs)

Generalities

In contact with the very reality of their roots, in a psychological operation campaign in guerrilla warfare, the comandantes will be able to obtain maximum psychological results from an Armed Propaganda program. This section is to inform the guerrilla student as to what Armed Propaganda Teams are in the environment of guerrilla warfare.

Combination: Political Awareness and Armed Propaganda

The Armed Propaganda Teams combine political awareness building with armed propaganda, which will be carried out by carefully selected guerrillas (preferably with experience in combat), for personal persuasion within the population.

The selection of the staff is more important than the training, because we cannot train guerrilla cadres just to show the sensations of ardor and fervor, which are essential for person-to-person persuasion. More important is the training of persons who are intellectually agile and developed.

An Armed Propaganda Team includes from 6 to 10 members; this number or a smaller number is ideal, since there is more camaraderie, solidarity and group spirit. The themes to deal with are assimilated more rapidly and the members react more rapidly to unforeseen situations.

In addition to the combination as armed propagandist-combatant each member of the team should be well prepared to carry out permanent person to-person communication, face-to-face.

The leader of the group should be the commando who is the most highly motivated politically and the most effective in face-to-face persuasion. The position, hierarchy or range will not be decisive for carrying out that function, but rather who is best qualified for communication with the people.

The source of basic recruitment for guerrilla cadres will be the same social groups of Nicaraguans to whom the psychological campaign is directed, such as peasants, students, professionals, housewives, etc. The campesinos (peasants) should be made to see that they do not have lands; the workers that the State is putting an end to factories and industries; the doctors, that they are being replaced by Cuban paramedics, and that as doctors they cannot practice their profession due to lack of medicines. A requirement for recruiting them will be their ability to express themselves in public.

The selection of the personnel is more important than the training. The political awareness-building and the individual capabilities of persuasion will be shown in the group discussions for motivation of the guerrilla as a propagandist-combatant chosen as cadres to organize them in teams, that is, those who have the greatest capacity for this work.

The training of guerrillas for Armed Propaganda Teams emphasizes the method and not the content. A two-week training period is sufficient if the recruitment is done in the form indicated. If a mistaken process of recruitment has been followed, however good the training provided, the individual chosen will not yield a very good result.

The training should be intensive for 14 days, through team discussions, alternating the person who leads the discussion among the members of the group.

The subjects to be dealt with will be the same, each day a different theme being presented, for a varied practice.

The themes should refer to the conditions of the place and the meaning that they have for the inhabitants of the locality, such as talking of crops, fertilizers, seeds, irrigation of crops, etc. They can also include the following topics:

Sawed wood, carpenters' tools for houses or other buildings.

Boats, roads, horses, oxen for transportation, fishing, agriculture.

Problems that they may have in the place with residents, offices of the regime, imposed visitors, etc.

Force labor, service in the militia.

Forced membership in Government groups, such as women's clubs, youth associations, workers' groups, etc.

Availability and prices of consumer articles and of basic needs in the grocery stores and shops of the place.

Characteristics of education in the public schools.

Anxiety of the people over the presence of Cuban teachers in the schools and the intrusion of politics, i.e. using them for political ends and not educational ones as should be.

Indignation over the lack of freedom of worship, and persecution, of which priests are victims; and over the participation of priests such as Escoto and Cardenal in the government, against the explicit orders of his Holiness, the Pope.

NOTE: Members of the team can develop other themes.

The target groups for the Armed Propaganda Teams are not

the persons with sophisticated political knowledge, but rather those whose opinion are formed from what they see and hear. The cadres should use persuasion to carry out their mission. Some of the persuasive methods that they can use are the following:

Interior Group/Exterior Group. It is a principle of psychology that we humans have the tendency to form personal associations from "we" and "the others," or "we" and "they", "friends" and "enemies," "fellow countrymen" and "foreigners," "mestizos" and "gringos."

The Armed Propaganda Team can use this principle in its activities, so that it is obvious that the "exterior" groups ("false" groups) are those of the regime, and that the "interior" groups ("true" groups) that fight for the people are the Freedom Commandos.

We should inculcate this in the people in a subtle manner so that these feelings seem to be born of themselves, spontaneously.

"Against" is much easier that "for." It is a principle of political science that it is easier to persuade the people to vote against something or someone than to persuade them to vote in favor of something or someone. Although currently the regime has not given the Nicaraguan people the opportunity to vote, it is known that the people will vote in opposition, so that the Armed Propaganda Teams can use this principle in favor of our insurrectional struggle. They should ensure that this campaign is directed specifically against the government or its sympathizers, since the people should have specific targets for their frustrations.

Primary Groups and Secondary Groups. Another principle of sociology is that we humans forge or change our opinions from two sources: primarily, through our association with our family, comrades, or intimate friends; and secondarily, through distant associations such as acquaintances in churches, clubs or committees, labor unions or governmental organizations. The Armed Propaganda Team cadres should join the first groups in order to persuade them to follow the policies of our movement, because it is from this type of group that the opinions or changes of opinion come.

Techniques of Persuasion in Talks or Speeches:

Be Simple and Concise. You should avoid the use of difficult words or expressions and prefer popular words and expressions, i.e. the language of the people. In dealing with a person you should make use of concise language, avoiding complicated words. It is important to remember that we use oratory to make our people understand the reason for our struggle, and not to show off our knowledge.

Use Lively and Realistic Examples. Avoid abstract concepts, such as are used in universities in the advanced years, and in place of them, give concrete examples such as children playing, horses galloping, birds in flight, etc.

Use Gestures to Communicate. Communication, in addition to being verbal, can be through gestures, such as using our hands expressively, back movements, facial expressions, focusing of our look and other aspects of "body language," projecting the individual personality in the message.

Use the Appropriate Tone of Voice. If, on addressing the people, you talk about happiness, a happy tone should be used. If you talk of something sad, the tone of the voice should be one of sadness; on talking of a heroic or brave act, the voice should be animated, etc.

Above All, Be Natural, Imitation of others should be avoided, since the people, especially simple people, easily distinguish a fake. The individual personality should be projected when addressing the population.

"Eyes and Ears" Within the Population

The amount of information for intelligence that will be generated by the deployment of the Armed Propaganda Teams will allow us to cover a large area with out commandos, who will become the eyes and ears of our movement within the population.

The combined reports of an Armed Propaganda Team will provide us with exact details on the enemy activities.

The intelligence information obtained by the Armed Propaganda Teams should be reported to the chiefs. However, it is necessary to emphasize that the first mission of the Armed Propaganda Teams is to carry out psychological operations, not to obtain data for intelligence.

Any intelligence report will be made through the outside contact of the Armed Propaganda Team, in order not to compromise the population.

The Armed Propaganda cadres are able to do what others in a guerrilla campaign cannot do: determine personally the development or deterioration of the popular support and the sympathy or hostility that the people feel toward our movement.

The Armed Propaganda Team program, in addition to being very effective psychologically, increases the guerrilla capacity in obtaining and using information.

In addition, the Armed Propaganda cadre will report to his superior the reaction of the people to the radio broadcasts, the insurrectional flyers, or any other means of propaganda of ours.

Expressions or gestures of the eyes, or face, the tone and strength of the voice, and the use of the appropriate words greatly affect the face-to-face persuasion of the people.

With the intelligence reports supplied by the Armed Propaganda Teams, the comandantes will be able to have exact knowledge of the popular support, which they will make use of in their operations.

4. Psychological Tactics, Maximum Flexibility

Psychological tactics will have the greatest flexibility within a general plan, permitting a continuous and immediate adjustment of the message, and ensuring that an impact is caused on the indicated target group at the moment in which it is the most susceptible.

Tactically, an Armed Propaganda Equipment program should cover the majority and if possible all of the operational area. The communities in which this propaganda is carried out should not necessarily form political units with an official nature. A complete understanding of their structure or organization is not necessary because the cadres will work by applying socio-political action and not academic theory.

The target populations of the Armed Propaganda Teams will be chosen for being part of the operational area, and not for their size or amount of land.

The objective should be the people and not the territorial area.

In this respect, each work team will be able to cover some six towns approximately, in order to develop popular support for our movement.

The Team should always move in a covert manner within the towns of their area.

They should vary their route radically, but not their itinerary,. This is so that the inhabitants who are cooperating will be dependent on their

itinerary, i.e., the hour in which they can frequently contact them to give them the information.

The danger of betrayal or an ambush can be neutralized by varying the itinerary a little, using different routes, as well as arriving or leaving without previous warning.

Whenever the surprise factor is used, vigilance should be kept in order to detect the possible presence of hostile elements.

No more than three consecutive days should be spent in a town.

The limit of three days has obvious tactical advantages, but it also has a psychological effect on the people, on seeing the team as a source of current and up-to-date information. Also, it can overexpose the target audience and cause a negative reaction.

Basic tactical precautions should be taken. This is necessary for greater effectiveness, as was indicated in dealing with the subject of "Armed Propaganda," and when it is carried out discreetly, it increases the respect of the people for the team and increases their credibility.

The basic procedures are: covert elements that carry out vigilance before and after the departure and in intervals. There should be two at least, and they should meet at a predetermined point upon a signal, or in view of any hostile action.

The team's goal is to motivate the entire population of a place, but to constantly remain aware that defined target groups exist within this general configuration of the public.

Although meetings may be held in the population, the cadres should recognize and keep in contact with the target groups, mixing with them before, during and after the meeting. The method for holding this type of meeting was included in the topic "Armed Propaganda," and will be covered in greater detail under the title "Control of Mass Meetings and Demonstrations."

The basic focus of the Armed Propaganda cadres should be on the residents of the town, where their knowledge as formers of opinion can be applied.

In the first visits of identification with the inhabitants, the guerrilla cadres will be courteous and humble. They can work in the fields or in any other form in which their abilities can contribute to the improvement of the living style of the inhabitants of the place, winning their trust and talking with them; helping to repair the fences of their cattle; the cleaning of the same, collaborating in the vaccination of their animals; teaching them to read, i.e., closely together in all the tasks of the peasant or the community.

In his free time, our guerrilla should mix in with the community groups and participate with them in pastoral activities, parties, birthdays, and even in wakes or burials of the members of said community; he will try to converse with both adults and adolescents. |He will try to penetrate to the heart of the family, in order to win the acceptance and trust of all of the residents of that sector.

The Armed Propaganda Team cadres will give ideological training, mixing these instructions with folkloric songs, and at the same time he will tell stories that have some attraction, making an effort to make them refer to heroic acts of our ancestors. He will also try to tell stories of heroism of our combatants in the present struggle so that listeners try to imitate them. It is important to let them know that there are other countries in the world where freedom and democracy cause those governing to be concerned over the well-being of their people, so that the children have medical care and free education; where also they are concerned that everyone have work and food, and all freedoms such as those of religion, association and expression; where the greatest objective of the government is to keep its people happy.

The cadres should not make mention of their political ideology during the first phase of identification with the people, and they should orient their talks to things that are pleasing to the peasants or the listeners, trying to be as simple as possible in order to be understood.

The tactical objectives for identification with the people are the following:

To establish tight relations through identification with the people through their very customs.

To determine the basic needs and desires of the different target groups.

To discover the weaknesses of the governmental control.

Little by little, to sow the seed of democratic revolution, in order to change the vices of the regime towards a new order of justice and collective well being.

In the motivation of the target groups, by the Armed Propaganda Teams, the cadre should apply themes of "true" groups and themes of "false" groups. The true group will correspond to the target group and the false one to the regime.

For the economic interest groups, such as small businessmen and farmers, it should be emphasized that their potential progress is "limited" by the government, that resources are scarcer and scarcer, the earnings/profits minimal, taxes high, etc. This can be applied to entrepreneurs of transportation and others.

For the elements ambitious for power and social positions, it will be emphasized that they will never be able to belong to the governmental social class, since they are hermetic in their circle of command. Example, the nine Government leaders do not allow other persons to participate in the government, and they hinder the development of the economic and social potential of those like him, who have desires of overcoming this, which is unjust and arbitrary.

Social and intellectual criticisms. They should be directed at the professionals, professors, teachers, priests, missionaries, students and others. Make them see that their writings, commentaries or conversations are censored, which does not make it possible to correct these problems.

Once the needs and frustrations of the target groups have been determined, the hostility of the people to the "false" groups will become more direct, against the current regime and its system of repression. The people will be made to see that once this system or structure has been eliminated, the cause of their frustration s would be eliminated and they would be able to fulfill their desires. It should be shown to the population that supporting the insurrection is really supporting their own desires, since the democratic movement is aimed at the elimination of these specific problems.

As a general rule, the Armed Propaganda teams should avoid participating in combat. However, if this is not possible, they should react as a guerrilla unit with tactics of "hit and run," causing the enemy the greatest amount of casualties with aggressive assault fire, recovering enemy weapons and withdrawing rapidly.

One exception to the rule to avoid combat will be when in the town they are challenged by hostile actions, whether by an individual or whether by a number of men of an enemy team.

The hostility of one or two men can be overcome by eliminating the enemy in a rapid and effective manner. This is the most common danger.

When the enemy is equal in the number of its forces, there should be an immediate retreat, and then the enemy should be ambushed or eliminated by means of sharpshooters.

In any of the cases, the Armed Propaganda Team cadres should not turn the town into a battleground. Generally, our guerrilla will be better armed, so that they will obtain greater respect from the population if they carry out appropriate maneuvers instead of endangering their lives, or even destroying their houses in an encounter with the enemy within the town.

A Comprehensive Team Program - Mobile Infrastructure

The psychological operations through the Armed Propaganda Teams include the infiltration of key guerrilla communicators (i.e., Armed Propaganda Team cadres) into the population of the country, instead of sending messages to them through outside sources, thus creating our "mobile infrastructure."

A "mobile infrastructure" is a cadre of our Armed Propaganda Team moving about, i.e., keeping in touch with six or more populations, from which his source of information will come; and at the same time it will serve so that at the appropriate time they will become integrated in the complete guerrilla movement.

In this way, an Armed Propaganda Team program in the operational area builds for our comandantes in the countryside constant source of data gathering (infrastructure) in all the area. It is also a means for developing or increasing popular support, for recruiting new members and for obtaining provisions.

In addition, an Armed Propaganda Team program allows the

expansion of the guerrilla movement, since they can penetrate areas that are not under the control of the combat units. In this way, through an exact evaluation of the combat units they will be able to plan their operations more precisely, since they will have certain knowledge of the existing conditions.

The comandantes will remember that this type of operation is similar to the Fifth Column, which was used in the first part of the Second World War, and which through infiltration and subversion tactics allowed the Germans to penetrate the target countries before the invasions. They managed to enter Poland, Belgium, Holland and France in a month, and Norway in a week. The effectiveness of this tactic has been clearly demonstrated in several wars and can be used effectively by the Freedom Commandos.

The activities of the Armed Propaganda Teams run some risks, but no more than any other guerrilla activity. However, the Armed Propaganda Teams are essential for the success of the struggle.

Conclusions

In the same way that the explorers are the "eyes and "ears" of a patrol, or of a column on the march, the Armed Propaganda Teams are also the source of information, the "antennas" of our movement, because they find and exploit the sociopolitical weaknesses in the target society, making possible a successful operation.

DEVELOPMENT AND CONTROL OF FRONT ORGANIZATIONS

Generalities

The development and control of front organizations (or "facade" organizations) is an essential process in the guerrilla effort to carry out the insurrection. That is, in truth, an aspect of urban guerrilla warfare, but it should advance parallel to the campaign in the rural area. This section has as its objective to give the guerrilla student an understanding of the development and control of front organizations in guerrilla warfare.

Initial Recruitment

The initial recruitment to the movement, if it is involuntary, will be carried out through several "private" consultations with a cadre (without his knowing that he is talking to a member of ours). Then, the recruit will be informed that he or she is already inside the movement, and he will be exposed to the police of the regime if he or she does not cooperate.

When the guerrillas carry out missions of armed propaganda and a program of regular visits to the towns by the Armed Propaganda Teams, these contacts will provide the commandos with the names and places of persons who can be recruited. The recruitment, which will be voluntary, is done through visits by guerrilla leaders or political cadres.

After a chain of voluntary recruitments has been developed, and the trustworthiness of the recruits has been established by their carrying out small missions, they will be instructed about increasing/widening the chain by recruiting in specific target groups, in accordance with the following procedure:

From among their acquaintances or through observation of the target groups - political parties, workers' unions, youth groups, agrarian associations, etc. - finding out the personal habits, preferences and biases, as well as the weaknesses of the "recruitable" individuals.

Make an approach through an acquaintance, and if possible, develop a friendship, attracting him through his preferences or weaknesses: it might be inviting him for lunch in the restaurant of his choice or having a drink in his favorite cantina or an invitation to dinner in the place he prefers.

Recruitment should follow one of the following guidelines:

If in an informal conversation the target seems susceptible to voluntary recruitment based on his beliefs and personal values, etc., the political cadre assigned to carry out the recruitments will be notified of this. The original contact will indicate to the cadre assigned, in detail, all he knows of the prospective recruit, and the style of persuasion to be used, introducing the two.

If the target does not seem to be susceptible to voluntary recruitment, meetings can be arranged which seem casual with the guerrilla leaders or with the political cadres (unknown by the target until that moment). The meetings will be held so that "other persons" know that the target is attending them, whether they see him arrive at a particular house, seated at the table in a particular bar or even seated on a park bench. The target, then, is faced with the fact of his participation in the insurrectional struggle and it will be indicated to him also that if he fails to cooperate or to carry out future orders, he will be subjected to reprisals by the police or soldiers of the regime.

The notification of the police, denouncing a target who does not want to join the guerrillas, can be carried out easily, when it becomes necessary, through a letter with false statements of citizens who are not implicated in the movement. Care should be taken that the person who recruited him covertly is not discovered.

With the carrying out of clandestine missions for the movement, the involvement and handing over of every recruit is done gradually on a wider and wider scale, and confidence increases. This should be a gradual process, in order to prevent confessions from fearful individuals who have been assigned very difficult or dangerous missions too early.

Using this recruitment technique, our guerrillas will be able to successfully infiltrate any key target group in the regime, in order to improve the internal control of the enemy structure.

Established Citizens, Subjective Internal Control

Established citizens, such as doctors, lawyers, businessmen, landholders, minor state officials, etc., will be recruited to the movement and used for subjective internal control of groups and associations to which they belong or may belong.

Once the recruitment/involvement has been brought about, and has progressed to the point that allows that specific instructions be given to internal cadres to begin to influence their groups, instructions will be given to them to carry out the following:

The process is simple and only requires a basic knowledge of the Socrates dialectic: that is the knowledge that is inherent to another person or the established position of a group, some theme, some word or some thought related to the objective of persuasion of the person in charge of our recruitment.

The cadre then must emphasize this theme, word or thought in the discussions or meetings of the target group, through a casual commentary, which improves the focus of other members of the group in relation to this. Specific examples are:

Economic interest groups are motivated by profit and generally feel that the system hinders the use of their capability in this effort in some way, taxes, import-export tariffs, transportation costs, etc. The cadre in charge will increase this feeling of frustration in later conversations.

Political aspirants, particularly if the are not successful, feel that the system discriminates against them unfairly, limiting their capabilities, because the regime does not allow elections. The cadres should focus political discussions towards this frustration.

Intellectual social critics (such as professors, teachers, priests, missionaries, etc.), generally feel that the government ignores their valid criticism or censors their comments unjustly, especially in a situation of revolution. This can easily be shown by the guerrilla cadre at meetings and discussions, to be an injustice of the system.

For all the target groups, after they have established frustrations, the hostility towards the obstacles to their aspirations will gradually become transferred to the current regime and its system of repression.

The guerrilla cadre moving among the target groups should always maintain a low profile, so that the development of hostile feelings towards the false Government regime seems to come spontaneously from the members of the group and not from suggestions of the cadres. This is internal subjective control.

Anti-governmental hostility should be generalized, and not necessarily in our favor. If a group develops a feeling in our favor, it can be utilized. But the main objective is to precondition the target groups for the fusion in mass organizations later in the operation, when other activities have been successfully undertaken.

Organizations of Cells for Security

Internal cadres of our movement should organize into cells of three persons, only one of them maintaining outside contact.

The cell of three persons is the basic element of the movement, with frequent meetings to receive orders and pass information to the cell leader. These meetings are also very important for mutually reinforcing the members of the cell, as well as their morale. They should exercise criticism of themselves on the realization or failures in carrying out individual subjective control missions.

The coordination of the three-member cell provides a security net for reciprocal communication, each member having contact with only an operational cell. The members will not reveal at the cell coordination meetings the identity of their contact in an operational cell; they will reveal only the nature of the activity in which the cell is involved, e.g., political party work, medical association work, etc.

There is no hierarchy in cells outside of an element of coordination, who is the leader, who will have direct but covert contact with our guerrilla comandante in the zone or operational area. The previous diagram does not indicate which new operational cell is the limit, but it indicates that for every three operational cells, we need a coordination cell.

Fusion in a "Cover" Organization

The fusion of organizations recognized by the government, such as associations and other groups, through internal subjective control, occurs in the final stages of the operation, in a tight connection with mass meetings.

When the guerrilla armed action has expanded sufficiently, armed propaganda missions will be carried out on a large scale: propaganda teams will have clearly developed open support of the institutions; the enemy system of target groups will be well infiltrated and preconditioned. At the point at which mass meetings are held, the internal cadres should begin discussions for the "fusion" of forces into an organization – this organization will be a "cover" source of our movement.

Any other target group will be aware that other groups are developing greater hostility to the government., the police and the traditional legal bases of authority. The guerrilla cadres in that group - for example, teachers - will cultivate this awareness-building, making comments such as So-and-so, who is a farmer, said that the members of his cooperative believe that the new economic policy is absurd, poorly planned and unfair to the farmers."

When the awareness-building is increased, in the sense that other groups feel hostility towards the regime, the group discussions are held openly and our movement will be able to receive reports that the majority of their operatives are united in common, greater hostility against the regime. This will be developed and the order to fuse/join will come about. The fusion into a "cover" front is carried out thusly:

Internal cadres of our movement will meet with people such as presidents, leaders, and others, at organized meetings chaired by the group chief of our movement. Two or three escorts can assist the guerrilla cadre if it becomes necessary.

Publish a joint communiqué on this meeting, announcing the creation of the "cover" front, including names and signatures of the participants, and names of the organizations that they represent.

After releasing this communiqué, mass meetings should be initiated, which should have as a goal the destruction of the Government control.

Conclusions

The development and control of the "cover" organizations in a guerrilla war will give our movement the ability to create the "whiplash" effect within the population, when the order for fusion is gives. When the infiltration and internal subjective control have been developed parallel with other guerrilla activities, a democratic guerrilla commander will literally be able to shake up the Government structure and replace it.

CONTROL OF MASS CONCENTRATIONS AND MEETINGS

Generalities

In the last stages of a guerrilla war, mass concentrations and meetings are a powerful psychological tool for carrying out the mission. This section has as its objective giving the guerrilla student training on techniques for controlling mass concentrations and meetings in guerrilla warfare.

Infiltration of Guerrilla Cadres

Infiltration of guerrilla cadres (whether a member of our movement or outside element) in workers' unions, student groups, peasant organizations, etc., preconditioning these groups for behavior within the masses, where they will have to carry proselytism for the instructional struggle in a clandestine manner.

Our psychological war team should prepare in advance a hostile mental attitude among the target groups so that at the decisive moment they can turn their furor into violence, demanding their rights that have been trampled upon by the regime.

These preconditioning campaigns must be aimed at the political parties, professional organizations, students, laborers, the masses of the unemployed, the ethnic minorities and any other sector of society that is vulnerable or recruitable; this also includes the popular masses and sympathizers of our movement.

The basic objective of a preconditioning campaign is to create a negative "image" of the common enemy, e.g.:

Describe the managers of collective government entities as trying to treat the staff the way "slave foremen" do.

The police mistreat the people like the Communist "Gestapo" does.

The government officials of National Reconstruction are puppets of Russian Cuban imperialism.

Our psychological war cadres will create compulsive obsessions of a temporary nature in places of public concentrations, constantly hammering away at the themes pointed out or desired, the same as in group gatherings; in informal conversations expressing discontent; in addition passing out brochures and flyers, and writing editorial articles both on the radio and in newspapers, focused on the intention of preparing the mind of the people of the decisive moment, which will erupt in general violence.

In order to facilitate the preconditioning of the masses, we should often use phrases to make the people see, such as:

The taxes that they pay the government do not benefit the people at all, but rather are uses as a form of exploitation in order to enrich those governing.

Make it plain to the people that they have become slaves, that they are being exploited by privileged military and political groups.

The foreign advisers and their counseling programs are in reality "interveners" in our homeland, who direct the exploitation of the nation in accordance with the objectives of the Russian and Cuban imperialists, in order to turn our people into slaves of the hammer and sickle.

Selection of Appropriate Slogans

The leaders of the guerrilla warfare classify their slogans in accordance with the circumstances with the aim of mobilizing the masses in a wide scale of activities and at the highest emotional level.

When the mass uprising is being developed, our covert cadres should make partial demands, initially demanding, e.g. "We want food," "We want freedom of worship," "We want union freedom" - steps that will lead us toward the realization of the goals of our movement, which are: GOD, HOMELAND and DEMOCRACY.

If a lack of organization and command is noted in the enemy authority, and the people find themselves in a state of exaltation, advantage can be taken of this circumstance so that our agitators will raise the tone of the rallying slogans, taking them to the most strident point.

If the masses are not emotionally exalted, our agitators will continue with the "partial" slogans, and the demands will be based on daily needs, chaining them to the goals of our movement.

An example of the need to give simple slogans is that few people think in terms of millions of cordobas, but any citizen, however humble he may be, understands that a pair of shoes is necessary. The goals of the movement are of an ideological nature, but our agitators must realize that food - "bread and butter," "the tortilla and red beans" - pull along the people, and it should be understood that this is their main mission.

Creation of Nuclei

This involves the mobilization of a specific number of agitators of the guerrilla organization of the place. This group will inevitably attract an equal number of curious persons who seek adventures and emotions, as well as those unhappy with the system of government. The guerrillas will attract sympathizers, discontented citizens as a consequence of the repression of the system. Each guerrilla sub-unit will be assigned specific tasks and missions that they should carry out.

Our cadres will be mobilized in the largest number possible, together with persons who have been affected by the Communist dictatorship, whether their possessions have been stolen from them, they have been incarcerated, or tortured, or suffered from any other type of aggression against them. They will be mobilized toward the areas where the hostile and criminal elements of the FSLN, CDS and others live, with an effort for them to be armed with clubs, iron rods, placards and if possible, small firearms, which they will carry hidden.

If possible, professional criminals will be hired to carry out specific selected "jobs."

Our agitators will visit the places where the unemployed meet, as well as the unemployment offices, in order to hire them for unspecified "jobs." The recruitment of these wage earners is necessary because a nucleus is created under absolute orders.

The designated cadres will arrange ahead of time the transportation of the participants, in order to take them to meeting places in private or public vehicles, boats or any other type of transportation.

Other cadres will be designated to design placards, flags and banners with different slogans or key words, whether they be partial, temporary or of the most radical type.

Other cadres will be designated to prepare flyers, posters, signs and pamphlets to make the concentration more noticeable. This material will contain instructions for the participants and will also serve against the regime.

Specific tasks will be assigned to others, in order to create a "martyr" for the cause, taking the demonstrators to a confrontation with the authorities, in order to bring about uprisings or shootings, which will cause the death of one or more persons, who would become the martyrs, a situation that should be made use of immediately against the regime, in order to create greater conflicts.

Ways to Lead an Uprising at Mass Meetings

It can be carried out by means of a small group of guerrillas infiltrated within the masses, who will have the mission of agitating, giving the impression that there are many of them and that they have popular backing. Using the tactics of a force of 200-300 agitators, a demonstration can be created in which 10,00-20,00 persons take part.

The agitation of the masses in a demonstration is carried out by means of sociopolitical objectives. In this action one or several people of our convert movement should take part, highly trained as mass agitators, involving innocent persons, in order to bring about an apparent spontaneous protest demonstration. They will lead all of the concentration to the end of it.

Outside Commando. This element stays out of all activity, located so that they can observe from where they are the development of the planned events. As a point of observation, they should look for the tower of a church, a high building, a high tree, the highest level of the stadium or an auditorium, or any other high place.

Inside Commando. This element will remain within the multitude. Great importance should be given to the protection of the leaders of these elements. Some placards or large allusive signs should be used to designate the Commando Posts and to provide signals to the sub-units. This element will avoid placing itself in places where fights or incidents come about after the beginning of the demonstration.

These key agitators of ours will remain within the multitude. The one responsible for this mission will assign ahead of time the agitators to remain near the placard that he will indicate to them, in order to give protection to the placard from any contrary element. In that way the commander will know where our agitators are, and will be able to send orders to change passwords or slogans, or any other unforeseen thing, and even eventually to incite violence if he desires it.

At this stage, once the key cadres have been dispersed, they should place themselves in visible places such as by signs, lampposts, and other places which stand out.

Our key agitators should avoid places of disturbances, once they have taken care of the beginning of the same.

Defense Posts. These elements will act as bodyguards in movement, forming a ring of protection for the chief, protecting him from the police and the army, or helping him to escape if it should be necessary. They should be highly disciplined and will react only upon a verbal order from the chief.

In case the chief participates in a religious concentration, a funeral or any other type of activity in which they have to behave in an organized fashion, the bodyguards will remain in the ranks very close to the chief or to the placard or banner carriers in order to give them full protection.

The participants in this mission should be guerrilla combatants in civilian clothes, or hired recruits who are sympathizers in our struggle and who are against the oppressive regime.

These members must have a high discipline and will use violence only on the verbal orders of the one in charge of them.

Messengers. They should remain near the leaders, transmitting orders between the inside and outside commandos. They will use communication radios, telephones, bicycles, motorcycles, cars, or move on foot or horseback, taking paths or trails to shorten distances. Adolescents (male and female) are ideal for this mission.

Shock Troops. These men should be equipped with weapons (Knives, razors, chains, clubs, bludgeons) and should march slightly behind the innocent and gullible participants. They should carry their weapons hidden. They will enter into action only as "reinforcements" if the guerrilla agitators are attacked by the police. They will enter the scene quickly, violently and by surprise, in order to distract the authorities, in this way making possible the withdrawal or rapid escape of the inside commando.

Carriers of Banners and Placards. The banners and placards used in demonstrations or concentrations will express the protests of the population, but when the concentration reaches its highest level of euphoria or popular discontent, our infiltrated persons will make use of the placards against the regime, which we manage to infiltrate in a hidden fashion, an don them slogans or key words will be expressed to the benefit of our cause. The one responsible for this mission will assign the agitators ahead of time to keep near the placard of any contrary element. In that way, the comandante will know where the agitators are, and will be able to send orders to change slogans and eventually to incite violence if he wishes.

Agitators of Rallying Cries and Applause. They will be trained with specific instructions to use tried rallying cries. They will be able to use phrase such as "WE ARE HUNGRY, WE WAND BREAD," and "WE DON'T WANT COMMUNISM." There work and their technique for agitating the masses is quite similar to those of the leaders of applause and slogans at the high school football or baseball games. The objective is to become more adept and not just to shout rallying cries.

Conclusions

In a revolutionary movement of guerrilla warfare, the mass concentrations and protest demonstrations are the principle essential
for the destruction of the enemy structures.

MASSIVE IN-DEPTH SUPPORT THROUGH PSYCHOLOGICAL OPERATIONS

Generalities

The separate coverage in these sections could leave the student with some doubts. Therefore, all sections are summarized here, in order to give a clearer picture of this book.

Motivation as Combatant-Propagandist

Every member of the struggle should know that his political mission is as important as, if not more important than, his tactical mission.

Armed Propaganda

Armed propaganda in small towns, rural villages, and city residential districts should give the impression that our weapons are not for exercising power over the people, but rather that the weapons are for protecting the people; that they are the power of the people against the FSLN government of oppression.

Armed Propaganda Teams

Armed Propaganda Teams will combine political awareness building and the ability to conduct propaganda for ends of personal persuasion, which will be carried out within the population.

Cover ("Facade") Organizations

The fusion of several organizations and associations recognized by the government, through internal subjective control, occurs in the final stages of the operation, in close cooperation with mass meetings.

Control of Mass Demonstrations

The mixture of elements of the struggle with participants in the demonstration will give the appearance of a spontaneous demonstration, lacking direction, which will be used by the agitators of the struggle to control the behavior of the masses.

Conclusion

Too often we see guerrilla warfare only from the point of view of combat actions. This view is erroneous and extremely dangerous. Combat actions are not the key to victory in guerrilla warfare but rather form part of one of the six basic efforts. There is no priority in any of the efforts, but rather they should progress in a parallel manner. The emphasis or exclusion of any of these efforts could bring about serious difficulties, and in the worst of cases, even failure. The history of revolutionary wars has shown this reality.

APPENDIX

The purpose of this appendix is to complement the guidelines and recommendations to the propagandist-guerrillas expressed under the topic of "Techniques of Persuasion in Talks and Speeches," to improve the ability to organize and express thoughts for those who wish to perfect their oratorical abilities. After all, oratory is one of the most valuable resources for exercising leadership. Oratory can be used, then, as an extraordinary political tool.

The Audience

Oratory is simultaneous communication par excellence, i.e., the orator and his audience share the same time and space. Therefore, every speech should be a different experience at "that" moment or particular situation which the audience is experiencing and which influences them. So the audience must be considered as "a state of mind." Happiness, sadness, anger, fear, etc., are states of mind that we must consider to exist in our audience, and it is the atmosphere that affects the target public.

The human being is made up of a mind and soul; he acts in accordance with his thoughts and sentiments and responds to stimuli of ideas and emotions. In that way there exist only two possible focuses in any plan, including speeches: the concrete, based on rational appeals, i.e., to thinking; and the idealized, with emotional appeals, i.e., to sentiment.

For his part the orator, although he must be sensitive to the existing mass sentiment, he must at the same time keep his cold judgment to be able to lead and control effectively the feelings of an audience. When in the oratorical momentum the antithesis between heart and brain comes about, judgment should always prevail, characteristic of a leader.

Political Oratory

Political oratory is one of the various forms of oratory, and it usually fulfills one of three objectives: to instruct, persuade, or move; and its method is reduced to urging (asking), ordering, questioning and responding.

Oratory is a quality so tied to political leadership that it can be said that the history of political orators is the political history of humanity, an affirmation upheld by names such as Cicero, Demosthenes, Danton, Mirabeau, Robespierre, Clemenceau, Lenin, Trotsky, Mussolini, Hitler, Roosevelt, etc.

Qualities in a Speech

In general terms, the most appreciated qualities of a speech, and specifically a political speech in the context of the psychological action of the armed struggle, are the following:
Be brief and concise
A length of five minutes [line missing in Spanish text]...that of the orator who said: "If you want a two-hour speech, I'll start right now; if you want a two-minute one, let me think awhile."
Centered on the theme
The speech should be structured by a set of organized ideas that converge on the theme. A good speech is expressed by concepts and not only with words.

Logic

The ideas presented should be logical and easily acceptable. Never challenge logic in the mind of the audience, since immediately the main thing is lost - credibility. As far as possible, it is recommended that all speeches be based on a syllogism, which the orator should adjust in his exposition. For example: "Those governing get rich and are thieves; the
Government officials have enriched themselves governing; then, the officials are thieves." This could be the point of a speech on the administrative corruption of the regime. When an idea or a set of guiding ideas do not exist in a speech, confusion and dispersion easily arise.

Structure of a Speech

Absolute improvisation does not exist in oratory. All orators have a "mental plan" that allows them to organize their ideas and concepts rapidly; with practice it is possible to come to do this in a few seconds, almost simultaneously with the expression of the word.
The elements that make up a speech are given below, in a structure that we recommend always putting into practice, to those who wish to more and more improve their oratorical abilities:
Introduction or Preamble
One enters into contact with the public, a personal introduc-

tion can be made or one of the movement to which we belong, the reason for our presence, etc. In these first seconds it is important to make an impact, attracting attention and provoking interest among the audience. For that purpose, there are resources such as beginning with a famous phrase or a previously prepared slogan, telling a dramatic or humorous story, etc.

Purpose or Enunciation

The subject to be dealt with is defined, explained as a whole or by parts.

Appraisal or Argumentation

Arguments are presented, EXACTLY IN THIS ORDER: First, the negative arguments, or against the thesis that is going to be upheld, and then the positive arguments, or favorable ones to our thesis, immediately adding proof or facts that sustain such arguments.

Recapitulation or Conclusion

A short summary is made and the conclusions of the speech are spelled out.

Exhortation

Action by the public is called for, i.e., they are asked in and almost energetic manner to do or not to do something.

Some Literary Resources

Although there exist typically oratorical devices of diction, in truth, oratory has taken from other literary genres a large number of devices, several of which often, in an unconscious manner, we use in our daily expressions and even in our speeches.

Below we enunciate many of their literary devices in frequent use in oratory, recommending to those interested moderate use of them, since an orator who over-uses the literary device loses authenticity and sounds untrue.

The devices that are used the most in oratory are those obtained through the repetition of words in particular periods of the speech, such as:

Anaphora, or repetition of a word at the beginning of each sentence, e.g., "Freedom for the poor, freedom for the rich, freedom for all." In the reiteration, repetition is of a complete sentence (slogan) insistently through the speech, e.g., "With God and patriotism we will overcome Communism because...:

Conversion is the repetition at the end of every phrase, e.g.: "Sandinismo tries to be about everyone, dominate everyone, command everyone, and as an absolute tyranny, do away with everyone."

In the emphasis, repetition is used at the beginning and at the end of the clause, e.g., "Who brought the Russian-Cuban intervention? The Government. And who is engaged in arms trafficking with the neighboring countries? The Government. And who is proclaiming to be in favor of nonintervention? The Government."

Reduplication, when the phrase begins with the same word that ends the previous one. For example: "We struggle for democracy, democracy and social justice." The concatenation is a chain made up of duplications. For example:

"Communism transmits the deception of the child to the young man, of the young man to the adult, and of the adult to the old man."

In the antithesis or word play, the same words are used with a different meaning to give an ingenious effect: e.g., "The greatest wealth of every human being is his own freedom, because slaves will always be poor but we poor can have the wealth of our freedom."

Similar cadences, through the use of verbs of the same tense and person, or nouns of the same number and case. For example: "Those of us who are struggling we will be marching because he who perseveres achieves, and he who gives up remains."

Use of synonyms, repetition of words with a similar meaning. For example: "We demand a Nicaragua for all, without exceptions, without omissions."

Among the figures of speech most used in oratory are:

Comparison or simile, which sets the relationship of similarity between two or more beings or things. For example: "Because we love Christ, we love his bishops and pastors," and "Free as a bird."

Antithesis, or the counter-position of words, ideas, or phrases of an opposite meaning. For example: "They promised freedom and gave slavery; that they would distribute the wealth and they have distributed poverty; that they would bring peace, and they have brought about war."

Among the logic figures are the following:

Concession, which is a skillful way to concede something to the adversary in order to better emphasize the inappropriate aspects, through the use of expressions such as: but, however, although, nevertheless, in spite of the fact that, etc. For example: "The mayor here has been honest, but he is not the one controlling all the money of the nation." It is an effective form of rebuttal when the opinion of the audience is not entirely ours.

Permission, in which one apparently accedes to something, when in reality it is rejected. For example: "Do not protest, but sabotage them." "Talk quietly, but tell it to everyone."

Prolepsis is an anticipated refutation. For example: "Some will think that they are only promises; they will say, others said the same thing, but no. We are different, we are Christians, we consider God a witness to our words."

Preterition is an artifice, pretending discretion when something is said with total clarity and indiscretion. For example: "If I were not obligated to keep military secrets, I would tell all of you of the large amount of armaments that we have so that you would feel even more confidence that our victory is assured."

Communication is a way to ask and give the answer to the same question. For example:

"If they show disrespect for the ministers of God, will they respect us, simple citizens? Never."

Rhetorical questions are a way in which one shows perplexity or inability to say something, only as an oratorical recourse. For example: "I am only a peasant and can tell you little. I know little and I will not be able to explain to you the complicated things of politics. Therefore, I talk to you with my heart, with my simple peasant's heart, as we all are."

Litotes is a form of meaning a lot by saying little. For example: "The nine commanders have stolen little, just the whole country."

Irony consists of getting across exactly the opposite of what one is saying. For example: "The divine mobs that threaten and kill, they are indeed Christians."

Amplification is presenting an idea from several angles. For example: "Political votes are the power of the people in a democracy. And economic votes are their power in the economy. Buying or not buying something, the majorities decide what should be produced. For something to be produced or to disappear. That is part of economic democracy."

The most usual plaintive figures of speech are:

Deprecation or entreaty to obtain something. For example: "Lord, free us from the yoke. Give us freedom."

Imprecation or threat, expressing a sentiment in view of the unjust or hopeless. For example: "Let there be a Homeland for all or let there be a Homeland for no one."

Commination, similar to the previous one, presents a bad wish for the rest. For example, "Let them drown in the abyss of their own corruption."

The apostrophe consists of addressing oneself towards something supernatural or inanimate as if it were a living being. For example: "Mountains of Nicaragua, make the seed of freedom grow."

Interrogation consists of asking a question of oneself, to give greater emphasis to what is expressed. It is different from communication, since it gives the answer and is of a logical and not a plaintive nature. For example: "If they have already injured the members of my family, my friends, my peasant brothers, do I have any path other than brandishing a weapon?"

Reticence consists of leaving a thought incomplete, intentionally, so that mentally the audience completes it. For example, "They promised political pluralism and gave totalitarianism. They promised political pluralism and gave totalitarianism. They promised social justice, and they have increased poverty. They offered freedom of thought, and they have given censorship. Now, what they promise the world are free elections..."

www.ingramcontent.com/pod-product-compliance
Lightning Source LLC
Chambersburg PA
CBHW050219270326
41914CB00003BA/474